AF228841

Jellyfish live in the sea.

Jellyfish are soft.

7

This jellyfish is pink!

9

This jellyfish is orange.

11

This jellyfish is red.

13

This jellyfish is see-through!

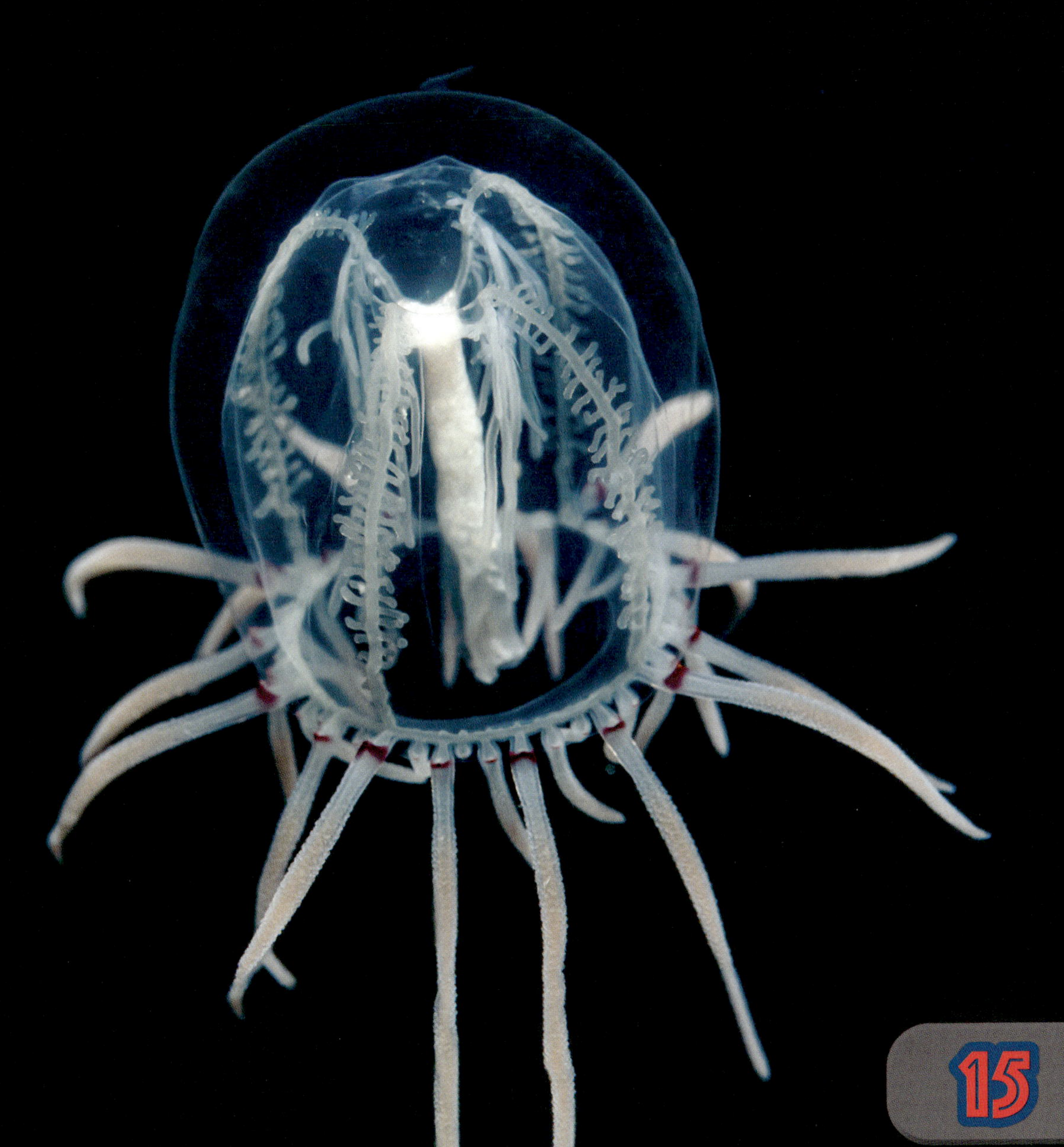

15

This jellyfish glows.

Look at jellyfish.

19

Don't touch jellyfish!

21

Jellyfish swim.

23

Published in 2026 by The Rosen Publishing Group, Inc.
2544 Clinton Street, Buffalo, NY 14224

First Edition

Editor: Theresa Emminizer
Book Design: Jeffrey Taylor

Photo Credits: Cover munawaekhan/Shutterstock.com; p. 3 Marina Ibrahem00/Shutterstock.com; p. 5 Muhammadsaifalkhan/Shutterstock.com; p. 7 Arina_B/Shutterstock.com; p. 9 LanKS/Shutterstock.com; p. 11 Ander5/Shutterstock.com; p. 13 Allexxandar/Shutterstock.com; p. 15 dfjac/Shutterstock.com; p. 17 Andrea Izzotti/Shutterstock.com; p. 19 Day 47 Media/Shutterstock.com; p. 21 Alexandree/Shutterstock.com; p. 23 bierchen/Shutterstock.com.

Cataloging-in-Publication Data

Names: Emminizer, Theresa.
Title: Jellyfish / Theresa Emminizer.
Description: Buffalo, New York : PowerKids Press, 2026. | Series: Superstars of the sea
Identifiers: ISBN 9781499451467 (pbk.) | ISBN 9781499451474 (library bound) | ISBN 9781499451481 (ebook)
Subjects: LCSH: Jellyfishes--Juvenile literature.
Classification: LCC QL377.S4 E46 2026 | DDC 593.5'3--dc23

Manufactured in the United States of America

CPSIA Compliance Information: Batch #CSPK26. For further information contact Rosen Publishing at 1-800-237-9932.

Find us on